Beyond Service
2024 Veterans' Voices Anthology

Compilation copyright @ 2024 Literary Cleveland

Literary Cleveland
13002 Larchmere Blvd.
Cleveland, OH 44120
litcleveland.org
info@litcleveland.org

ISNB: 978-1-7345589-1-3

All rights reserved. No part of this publication may be reproduced, distributed, or transmitted in any form or by any means, including photocopying, recording, or other electronic or mechanical methods, without the prior written permission of the publisher, except in the case of brief quotations embodied in critical reviews and certain other noncommercial uses permitted by copyright law. For permission requests, write to the publisher at the address above.

Any views, findings, conclusions or recommendations expressed in this program do not necessarily represent those of Literary Cleveland, the VA Northeast Ohio Healthcare System or other supporting organizations and funders.

Book cover by Each + Every
Book interior layout and printing by Outlandish Press
Produced by Literary Cleveland
Edited by Christopher Johnston

Introduction
Veterans' Voices instructor Christopher Johnston

After they leave military service, veterans often find themselves on a long and sometimes challenging journey. Fortunately, programs such as Literary Cleveland's Veterans Voices provide a productive, positive outlet either for their pent-up frustrations or for their pent-up creative gifts and inclinations. We've found many veterans possess creativity in abundance, but they don't always have the opportunity for creative expression.

As one of the instructors, along with Army National Guard veteran Mansa Bey, it has been my absolute pleasure and honor to witness the wide range of writing skills and creative ideas that our veteran writers displayed this past year in the Veterans Voices workshop sessions week after week. Many wanted to memorialize their diverse experiences in the service or honor their fellow veterans or lost comrades in arms. A few chose to use writing to exorcise some of the more difficult aspects of being in the military.

Several of our writers opted to write a poem that captures their military experiences. Some chose straight-up memoir pieces. A couple desired to craft a science fiction story that drew on their extensive knowledge of technology, aviation, naval or infantry techniques, or the world of modern warfare. These writings usually take the reader on colorful, detail-laden trips into the veteran writers' treasure-trove of memories or dreams or fantasies. Several of our writers have focused on their post-service experiences in their lives that have little or nothing to do with their time in the military.

Moreover, we know the healing powers of creativity. We understand the great joy that comes from writing a piece that you know will be collected for family, friends and posterity to read in a Veterans Voices anthology. We know the satisfaction that comes from having your work recognized and published. We know the fulfillment that accompanies having your work featured in the same collection as that of your fellow veterans and participants in the workshop where you have enjoyed the weekly camaraderie of the creative writing arena.

It has certainly been a delight to go along with them for the ride on their diverse and far-reaching creative expeditions into the realms of memory or imagination. It has also always been a tremendous honor to moderate in the workshop with these men and women who have served their country so ably, diligently and valiantly. We know you will enjoy venturing into their fertile and well-crafted memoir and creative pieces, too.

We thank our veterans for their service. We thank you for supporting their writings and Literary Cleveland.

CONTENTS

Ethiopian Encounter
    Richard Asbury...7
Goodbye—Shakedown
    Richard Asbury...11
Respect of Service
    Kevin Chapman..14
Sometimes There Are a Few Good Men
    Michael T. Conway..19
Mini Memories 16
    Cheryl Darby..24
Pool of Blue Torture
    Cheryl Darby..27
The Hankie
    Paul Facinelli..29
Camp David Mom
    Debra Gipson...31
Another Day in the Desert
    Brent Herzberger...34
Chappie
    Brent Herzberger...36
Oh, So Long
    Dennis J. Jackson..37
The Boys at the Wire
    Douglas Kulow..39
On the Other Hand
    Don Lee..42
Movement to Contact: Day One.
    Stephen May..43

CONTENTS

Vendetta
    Raye Oheidhin..47
What Color Is Your Love?
    Kevin D. Smith...48
Invisible
    Holly Spohn..49
We'll Always Have Paris
    Jaelle Terrell...54
The Courage of Spirit
    Callie Travers...59
Frosted Glass
    Vickie Williams..63
Thank You for Your Service, Fool!
    Ken Woolfolk..66
Contributors..69
Acknowledgements...73
About Literary Cleveland..75

Ethiopian Encounter
Richard Asbury

It was a bright, clear day in the fall of 1973. The air was crisp, with a slight breeze that tickled my cheeks. I was in Athens, Greece. My "Steaming Buddy" (John Henry Broomfield) and I flagged down one of the ubiquitous, dangerously driving cabs. I enjoyed barking out our desired destination in my very limited Greek. "Placa Sinta. Gregora. Parakaló." (Constitution Square. Hurry. Please.)

Constitution Square was the center of Athens back then. The Changing of the Guard Ceremony was held in front of the Parliament building and other government buildings bounded it on three sides. The fourth side was the famous St. George Hotel. That hotel hosted heads of state and dignitaries galore over the years. I was told that local small businessmen bid for the job of doorman each year. Why? To see and to be seen, as well as to hear the high, and often hidden, verbal exchanges by the famous visitors. They also hoped to earn "baksheesh" (bribes) from photographers who wanted information as to the expected comings and goings of celebrities and dignitaries.

"Broom" and I exited a cab in front of our favorite coffee shop that was tucked into the corner of Constitution Square. It was chilly. In the summer, we sat and sipped "al fresco." Now, we found seats near the small heaters that we noticed were placed in the glassed-in lounge. Broom ordered his habitual, "Nescafé, black," which was served in a glass mug with a globe painted on it. I ordered my "café au lait avec sucre" (espresso coffee, cream and sugar).

Then, the door at the back of the lounge opened allowing a sharp draft to blow in. We turned and saw two African couples walk in and choose seats near the heat, and us. I said to Broom, "They look Senegalese, to me." The men were average height, and their curly hair was trimmed into neat Afros. One man had a dark mocha hue, with bright eyes and a smile that revealed brilliant white teeth. The other man was a bit taller, caramel-colored, and appeared a bit older than the rest of the group.

And ... ah, the women! The one that first caught my eyes was the color of coffee, with a touch of copper that made her skin appear like a smooth, worn penny. She had a rhythmic stride and impish cheeks. She was petite but had well-formed calves that showed below her winter garb and hips that adequately filled her coat. The other woman was taller, slimmer, and had a stately look about her. She also appeared to be a bit older than her girlish-looking friend. She had a slight amber tint to her smooth, shiny skin.

Broom and I were mesmerized. We had been at sea for 30 days. Seeing beautiful, elegant, Black women, even if they were escorted, perked our blood, a little. I listened to them speak It sounded like Arabic. Which I could not speak at all beyond, "as-salaam alaikum." I said to Broom, "Let me see if they speak French," which I had had in school, but not at the fluent level. I leaned across the aisle that separated us. I asked the little cute one if she spoke French. She smiled and said, "Oui."

We carried on a conversation in very basic French. I learned that they were Ethiopians. Her name was Yeshi and her girlfriend's name was Almaz. They were in Athens studying pharmacy. They had to learn Greek for their first year, then in their second year, they would learn the science of drugs in earnest. The two gentlemen were their cousins, Abebe and Dula, who worked on a cruise ship home-ported in Athens.

I stopped and stuttered for a few moments. I smiled at Yeshi, then turned and said to Broom," Man, I'm running out of vocabulary!" Yeshi smiled and said sweetly, "Would you like to speak English now?" Everyone laughed heartily. I was a bit embarrassed and said, "Absolutely!"

We began a banter among us that yielded interesting and enlightening facts all around.

We learned:
- Yeshi's father was the mayor of Addis Ababa, the capital of Ethiopia.
- They were of two different ethnic groups. Yeshi was Amhara. Her people were directly related to the emperor. They lived in the northern central region of the country in the mountainous area. Her family lived in Addis, the capital.
- Almaz was Eritrean. She lived in Asmara in the northeast, near the Red Sea.
- There were many Italians and Greek businessmen in their cities.
- Ethiopian spaghetti was a common dish. Very spicey, but still very Italian.
- They were educated in English from a young age.
- They learned that Broom was a 15-year Navy veteran, and I was less than a two-year vet.
- They saw the USA as an ally to be emulated in certain ways.
- The ladies were amazed at the size of our ship and that it flew jet planes.

I loved to hear them speak Amhara, their national language. It had a soft, melodic tone, with a flowing rhythm to it. Sort of like French, yet not so nasal. When they shifted to English, some of their words had a pronounced British sound. Our lively discussions went on for more than an hour. Then Abebe announced that he and Dula had to head back to the ship to serve the evening meal to the crew. Almaz reminded Yeshi, who had been in constant chatter with everyone, that they too had to return to their dorm or they might miss their evening meal.

We said our good-byes, captured their phone number, and exchanged addresses. We promised to meet again when Forrestal was in port.

As we rode to our Shore Patrol station, we excitedly reviewed the encounter.

"Really some class ladies," I said. "That's for sure. So poised and lady-like," said Broom.

"Did you look at those eyes? Yeshi had eyes as big as hubcaps!" I said gushingly.

"Yeah, and Almaz had little green flecks in her brown eyes," Broom said through a broad smile.

Yeshi had excited my young heart. She inspired me to write a poem.

Goodbye—Shakedown
Richard Asbury

My week in Ethiopia was ending. I had planned to get to intimately know the Ethiopian girl,
Yeshi Mulageta, whom I met in Athens, Greece, a year earlier. We had dated every time my ship was in Athens and we had exchanged letters frequently. I was becoming enamored with her. I wanted to see her country, meet her family, understand her life. Maybe marry her.

In June of 1973, she invited me to her country, if I could, in the summer of 1974 when she was home from school in Athens.

I was tremendously excited! I planned the trip for a year. We exchanged letters regularly until May 1974. Then things went silent ... What was going on? I had my leave approved. I had bought my ticket. I had my traveler's checks (this was before Visa cards). I had decided on what I wanted to do. I was charged up and ready to go.

Then, it happened in June 1974. Emperor Haile Selassie was overthrown in a bloodless coup. Yeshi had fled the country. I would learn that her father had been the last mayor of Addis Ababa before the coup. His whole family had to flee while he remained in Addis to turn over the city to the rebels.

I had given Yeshi my flight arrival information and she had a picture of me. Her cousin,
Bucrazia, met me at the airport and said he was to be my guide. His only fee was to pay his
meals and transportation for each day he accompanied me. I learned that the average young Ethiopian loved James Brown's "I'm Black and Proud" and the hopes and dreams of African Americans.

Most spoke fluent English and were curious about Black life in America – it was a week I will never forget.

Around noon on my fifth day in Addis, I went to the front desk after lunch. The clerk said I had an important-looking letter – but no note from Kate or Yeshi. I had told both of them, during our conversations, that they could write me at the fleet post office address, which was: USS FORRESTAL, my name, HRDC, FPO New York City NY.

I opened the "important" looking letter. It was from the U.S. Consulate. It advised me to come to the consulate and get an exit visa. The political situation had changed and as a military member, I should return to my military base as soon as possible. I immediately caught a cab to the consulate. They stamped my travel orders and I was given an exit visa. I went back to the hotel, packed, and was able to be scheduled for the first flight that next morning.

***

I arrived at the airport bright and early. After checking in at the Ethiopian Air desk, I headed for my gate. While walking to my gate, I was confronted by two semi-official-looking soldiers. They were carrying Kalashnikov rifles on their shoulders and Makarov pistols on their hips. They were, I assumed, part of the interim army after the overthrow of the emperor. They were tall, lean, keen-featured, and dark-complected, with the silky, semi-straight hair of many Ethiopians. The smaller of the two said in heavily accented English, "Where you go, brother?" "To Greece," I replied. "Let me see your papers," he snapped. For the first time in my life, I was anxious, bordering on terrified.

"Oh, you have no passport. Oh, Oh, … You U.S. Military. Right?

You need no passport?" What part your military you work in?" "Navy...ships...large boats," I replied.

"What you doing here in Addis?" he asked. "Visiting friends," I said nervously. He turned and
said something to his comrade in Amharic. Then he said, "You cannot leave the country with cash. Give us all your cash in our money and yours." I clumsily gave him all I had in my wallet while silently thanking God that I had stowed $100 U.S. in my suitcase. The taller one said, "Give me the watch." I complied. He looked it over, scornfully. It was only a Seiko that I had bought at the base exchange in Norfolk for $100 or so. The shorter one talked to his partner in Amhara. Their conversation became a heated exchange ... I began to worry... Was I going to become a prisoner of their new government? Was I going to be accused of being a drug smuggler? I had no contacts beyond the embassy and Yeshi. She had gone to Italy with her extended family. I didn't know where Bucrazia, my guide, lived either! I was totally alone.

My two captors lowered their voices and the shorter one said, "You are a Black American
brother. You are a soldier, like us. You have orders to return to your boat... Go... Get on your
plane."

My heart was lifted through the roof. I walked normally until my "inspectors' were out of sight; then I sprinted to my gate on my strong, 24-year-old legs. They had almost given my seat away. But my U.S. Consulate stamp, on my military leave orders and I.D. saved me again.

Thank God for Uncle Sam!!!

Respect of Service
Kevin Chapman

"Thank you for your service."

There is a deep sense of imposter syndrome that nags at me, and many other veterans, every time I hear this.

We'll hear it for a variety of reasons. Usually, it's said with some sort of reverence that I don't feel myself for my time in uniform.

My go-to response is "I got more out of the Navy than they ever got out of me," and that's the God's honest truth.

In the Navy, I got the high school shyness knocked out of me and replaced with an extroverted outgoingness that I now have to rein in.

I developed a sense of adventure as I saw the world. The Artic Circle, Germany, France, Israel, Egypt, Kuwait, Iraq, Yemen, Bahrain, Saudi Arabia, St. Thomas, the list goes on.

I wouldn't have my two wonderful sons, Jacob and Nicholas, because fate wouldn't have improbably crossed my path with Teresa, my beautiful wife of thirty-two years.

That wonderful lady who tries to get me to see the flip side of the coin. "You served. You took the risks. You got the lumps, earned every number of that disability rating from the VA."

When she tells me this, I agree it is most certainly true, except you can't play the lottery without buying a ticket.

Shrugging my shoulders when I admit, "I mean, yeah, I did serve." No one made me join the Navy though. I did that on my own, signing up at sixteen. My eighteenth birthday spent in boot camp.

How many would do that without the carrot of benefits the recruiting officer dangled in front of us?

The advantage I most used was the training I was given. That boost when applying and testing for civilian jobs.

There are no frivolous, easy schools in the military. At least not for me, as I was pushed to exceed and succeed.

I started my journey with Interior Communication Electricians, a school in San Diego, before I headed to my first ship, the USS Theodore Roosevelt in Norfolk.

Before I headed to the Persian Gulf for duty, the USS LaSalle sent me to CCTV school.

That's how it went for each successive duty station. One more school under my belt.

I can hold my head up for my performance in all of them. I had to study, and I had to put in effort.

I look back and laugh at how that training was encouraged.

There was one A-school instructor, who shall remain nameless, he sat me down for a full day outside of class and helped me learn calculus by literally smacking me upside the back of my head EVERY ... SINGLE ... TIME ... I got an answer wrong. Call it what you will.

It worked.

A few years later that same instructor ended up as a student in the same class as me in Fresnel Lens school.

He didn't remember me. Smacking me was not a singular teaching event for him.

The time he and the other instructors invested in me is still deeply appreciated.

Schooling aside, did I spend over a year in a combat zone and have some truly scary moments? Yes.

Did I get that all so precious Combat Action Ribbon? Yes.

Do I have some mental bumps and bruises from oil well fires, minesweeping ops and a surprise pirate attack? Yes.

But to me, it wasn't anything special. Nothing anyone else couldn't do. I was challenged and I overcame, but did I suffer?

Comparatively, no.

Yes, I rightly earned those brightly colored ribbons and doodads that sit on my shelf, or as one mentor I had called them, "just another hole in your uniform."

Yet, that military mindset of "embracing the suck" makes accepting a "thank you" feel like I am weaponizing gratitude.

It was pointed out to me in therapy that I ate and drank my feelings. But everyone does that in the military, don't they? It's the

American way to push down the bad with a burger and a beer.

I had a lot of good times that way. I made friends I still have almost forty years later.

Benefits were reaped without the heavier cost that some of my brothers- and sisters-in-arms paid.

I am listed as a Combat Veteran. However, there is a deep-seated feeling in my gut that I haven't earned the honor to say that.

That after my time in, I got off cheap when it comes to the cost of life, liberty, and the pursuit of civilian happiness. My time in service came without the crushing moral injury others experienced. My sense of right and wrong towards justice is still intact.

When my youngest son Nick was still young, we would go to events and sometimes there would be a call for all veterans to stand up. Sometimes I would get up, but I stopped because it didn't sit right with me. In my soul, I feel I could have, should have, done more, even if I don't know what.

One day Nick asked why I was still seated. When I told him the honest reason of why, he said, "Can I stand for you?" How can anyone say "No" to that. It made my heart feel good.

Other veterans though? I will go far out of my way to stand for them, support them, their businesses, and charities. Especially the charities.

We've all heard the horror stories. We've met Soldiers, Sailors, Marines and Airmen who weren't so lucky. People who got chewed up, spit out, and left behind on the ground.

To combat that, my church sends out cards to active-duty military.

Because of how we met when I was in, my wife and I have sent care packages to military aid groups.

Occasionally we'll hear back from people who feel alone, have cried themselves to sleep as friends, family and lovers have moved on without them.

The opposite of paranoia is pronoia, the belief that someone or some entity is secretly working, not against you, but for your benefit.

Whether I credit it to God or fate and chance, as I contemplate my retirement this year from my civilian job, it does feel like I have had a hand on my shoulder.

I was guided down a better path than the one I would have followed if I had stayed in that small Missouri town I grew up in.

So, thanking me? It's not something I feel I deserve. But I will take it in honor for those who do.

Sometimes There Are a Few Good Men
Michael T. Conway

The one thing you can't do with trauma is reach for a button and shut it off. You have to find a way to let it go. But you can't. If the trauma touched your body, then every time you feel your body you remember the trauma. Like a rope burn. Or any burn for that matter.

Maybe you are watching a movie and see a scene that takes you back. You flash back and then sometimes you stay back there, in that black space with the bright light that you can't close your eyes to. You want to close your eyes, but it makes no difference if you did. You want to cry, but nothing comes out. No one is coming to save you. You are helpless. You don't want to see what else is there, waiting for you to show up. No one is there to talk to. No one understands. Hearing a melodramatic song on the radio breaks your mind into pieces. Trauma is lonely. Writing this story is lonely and keys my trauma. But I will keep it in check until I reach for that glass one more time tonight.

Loneliness is the cruelest thing you can do to a human being because you let that person know they are banished, not wanted, unacceptable, not worthy of living. And don't you dare tell me to turn the other cheek. I did that a lot trying to make things work and got slapped so hard again I got a concussion. The only way you can help me is a way I can't really explain. Maybe you letting me know I am being believed is the starting place. I need to have a vision of you holding my hand, or better yet, a lovely woman behind me, rubbing my shoulders in the waning autumn sunlight, as we wear our white V-neck sweaters and lean back against a fruit tree overlooking the faraway valley. We are watching the fog rolling in with her whispering something in my ears I can't make out, don't care to know, because I need something I can feel that seems real. I need to believe I am wanted when I know I am not. I can't be lonely, not another single day. And at least today is almost over.

I can't explain everything, and I want to, but I don't, but I will. I am from a culture you can't possibly understand but you can let me know you will try and that means a lot to me. There are rules for warriors. One of them is you don't question the rules. What is normal for a veteran is not normal for a civilian. Killing is reasonable conduct for a warrior. It is not for a civilian. Risking your life for the greater good is normal for a warrior. It is not for a civilian. So, I write to manage my fear of having to feel things. Writing about what happened to my clients in my law practice after serving years in the Marines gives me a sense of control. I also want to teach.

There is a saying that there is only one real story: the story between good and evil. Maybe. But that story is relative to finding the truth just like anything else. Everyone wants to win their case in court because we believe we deserve to win because we are just, and the other side is evil. Why else would we be here?

Enough about me. I was wondering, do you trust the cops? Answering is a yes- or no-type-deal. Suddenly, a Blackhawk-style helicopter bursts out of the overcast sky heading north, its rotors blasting me up and off my ass. He's headed where I'm headed. Up the highway. My daydreaming is over for today. Driving. Thinking about my client Avery.

Avery likes the idea of being a cop maybe because she likes the idea of the law being something noble to serve and that doing so makes her a respected person. But the superiors she worked for made her pay for that dream by convincing her not to respect herself in exchange for career security. She bought into the idea that letting the boys have a little taste is not a price that is too high. When they wanted the whole dinner, she did not know what to do but gave it to the guy who could protect her from the rest. She was talked out of herself by people in power over her. That uniform meant nothing to others who wore the same uniform. It was just a costume to them. It was always an idea that stood for something else to Avery.

There are some things that are so common they do not generate TV news. The conversion to Christianity, the marriage ending in divorce, the sexual harassment of a woman by the men she works with. Maybe sexual harassment is too common. Maybe men and women cannot work together. Working with people who carry guns requires trust. I mean real watch-your-back trust. So, when you can no longer trust people with your life, what do you do next? To protect yourself you stand off alone, thinking that way you would be safer.

But are you really? That standing off has a price called loneliness. Loneliness is the cruelest thing you can do to another human being. Avery was watching the wiper blades occasionally beating back against the drizzle. Cars rolled by slowly on the interstate, splashing melt-water spray into the night air. Headlights glared and the occasional tractor trailer slowed down making a lot of noise with its exhaust stacks. She pondered questions. What is it that makes you vulnerable to the dishonest, con artists, the abusers and the incompetent idiots with fingers on triggers? What is fear of death? You swore to give your life. Do you have the right to be afraid? Do you have to take it for the team? Do you have the guts to stare into the black holes in the face of a skull? The dark inside there is as empty as any space that the government has on its employee roster after you are killed in action.

Or is death something else? Is it fear of punishment in hell or fear of not being remembered as counting for something while you were here? Does it matter that you were here at all? You serve, so it can't matter to yourself. It can only matter to other people who relied on you. Death and loneliness have one thing in common: it is as if you never existed.

Driving back to the station, the occasional car threw more spray onto the windshield momentarily blurring the view of two red taillights in front of her. No one was speeding. Then that car slipped off the road and disappeared. Wow, oh my God, and then some. Avery fired up the roof on her cruiser and called dispatch. The nighttime world was now lit by a blue and white strobe dancing off the rain and the road. Parking, then running to the edge of the highway, she could see taillights in the air below her and hear the sound of a barking dog. A woman was crying. The smell of stale beer wafted through the dense air. She knew this woman from before somewhere. It was high school she thought. Look what became of her. She was that girl you knew but never talked to unless you had to.

Avery was trying to pry open the jammed door to no effect. The engine was still running and carbon monoxide was filling the car's passenger area. Car exhaust makes you sick. Really sick. Avery, using her night stick, smashed the window to get some air in there. Window glass was splintered, but not broken in the front windshield. It looked like a Coke bottle dropped on a concrete floor except that you could see an outline of the woman's nose and jaw pressed into the glass. No front airbag went off. That's odd. Little bluish-green bits of glass arranged in a broken spray that stayed in one place at the same time except for the molded face in the glass, staring forward into the rainy night in pain. Speaking into her mic on her vest Avery called dispatch to send EMS and Jaws of Life because the door frame and roof of the car were welded together by the force of impact. It was impossible for her to extract the driver.

And it was too late. The driver slumped over the wheel in her own vomit, and the dog stopped barking. The rain fell harder now and had some ice mixed in it. That was nightshift. Bring a recovery team while you are at it. 10-4. The radio hissed. 0233 check. All you can do is stand there with your uniform on, in the pouring rain, and wonder if she had someone you needed to call and tell. 0250 check.

Mini Memories
Cheryl Darby

I lost my maternal grandfather shortly after joining the military and arriving in Orlando, Florida, for bootcamp. My mother made the decision to convince me not to come home and attend the funeral, because she knew that if I returned home, I would not return to the military. I believe the previous calls home and my tears from missing my family and my sharing with her the early mornings that consisted of marching in a type of heat I was not familiar with affected her, until my adventures in boot camp helped her make the decision to not insist that I come home and pay my respects for my grandfather. Sweat would pour from those uniforms that made me look like I was in prison. Gnats were drawn to the wet, sweaty dungarees that caused me to swat constantly at these pests unless I was in formation. I gained unplanned weight and a sunbaked tan.

Teamwork was important in our company and we had to learn to work together. We learned to march to the galley and eat our chow in a limited amount of time. There wasn't time to lollygag if you wanted to eat because there weren't opportunities to eat outside of the assigned time. Learning how to fold our uniforms and place them inside of our seabags, which was a large green bag with straps, made you learn how to balance the bag in the center of your back.

Mail call was always a joy for me. Of course, to others it was a sad experience. It was always nice to share the contents of my letters with shipmates who didn't receive mail. Listening to music was a good pastime as well. I really didn't care who was listening even if I sang off-key "Out on a Limb" by Teena Marie and other songs that made me happy. We enjoyed one particular shipmate that would add some rhythm to cadence. I was even given the opportunity to lead cadence a few times and, boy, did I enjoy.

Sometimes I would irritate my bunkmate since I was assigned the top rack. As she read her letter with a small flashlight in the privacy of her bunk, I would hang my head down until she would realize I was peering down over her as if I could read her letter. I honestly couldn't see the words, but judging by the large smile on her face the letter was from someone that was able to bring a smile to her face.

There were the inspections that would have us on edge if our uniforms and other laundry wasn't folded in the correct width and length. There were always a few shipmates who had military family members and were able to pass those inspections with flying colors and they had no cares in the world. Then there was those who were Nervous Nancys who were on pins and needles until the inspections were completed. Some of the major tasks that we put into action was the shining of our boon dockers and ensuring that our gig lines were lined up properly. Those that were pros would come assist those of us who were either struggling or were moving a little too slow. They truly took teamwork to heart. One evening, or shall I say the wee hours of the morning, I had to go to the head and as I walked inside, I noticed a large palmetto bug on the ceiling. I kept walking and sat on the toilet seat. I looked back up at the ceiling and didn't see the bug anymore.

As I sat in relaxed mode, I felt my hair rising up and moving. When I realized that my hair had become its nest, I began to frantically pull my hair and scream at the top of my lungs, waking up a large majority of my shipmates. And they weren't too happy. That was when I learned the palmetto bug had wings. The palmetto was able to fly away safely without me crushing him in the tangled wool nest. So, to this day if I see I bug on any ceiling and then it disappears, I begin to nervously check my hair. I refuse to be a moving nest transporting a palmetto to their place to bask in the sun.

Pool of Blue Torture
Cheryl Darby

Commercials can influence you either negatively or positively. And so, it was the case with the commercial for joining the military. My recruiter was able to seal the deal with convincing me on what I would experience. The question was the career choice that I had in mind. I didn't have a clue, so he said, "Would you like to be on the radio and television like Johnny Carson?" My answer was "yes." So, my choice was to pursue the career of a Radioman.

I recently came across my bootcamp yearbook and bootcamp memories took me back to those days. To my surprise, inside of the book was the original words of encouragement from my shipmates of company K108 for me to pass the swim test. I remember when K108 went for our swim test. I watched as each seaman recruit would climb the ladder to jump off the diving board into what I saw as the blue endless waters.

For those who didn't have the nerve to take the leap of faith into the water, they were ordered to press their face up against the wall and have their back toward everyone within eyesight. To me, they looked so ridiculous, there was no way on earth I was going to follow suit and do the same. So, when my turn came to jump, I jumped. We were warned that the lifeguard who was in the water was there to save us and that we were not to grab him.

Now, hearing those instructions was easier than following them. When I hit the water, I
panicked and grabbed the lifeguard. And he dunked my head under the water. As my head
reached above the water again, I panicked and grabbed his arm,

and once again I was dunked under water. After several times of getting dunked back under water and swallowing water, I finally realized that if I wanted to get out of the pool of blue torture, I would have to comply to the rules and, after a few minutes of floating, I was finally able to vacate my pool of blue torture.

Once I returned to the barracks, my shipmates had letters wishing me a Happy Birthday
and congratulating me on finally passing my swim test. Yes, my fellow veterans, I passed my
swimming test on my 19th birthday. Reading the letters from so many years ago, brought back so many memories that I will carry with me for the rest of my life.

The Hankie
Paul Facinelli

My wife is married to a different man now, six-two, with the physique of a cornstalk, 175 with a roll of quarters in his pocket.

Whenever I imagined whom my wife might marry, I conjured a professorial type, facial hair, two elbow patches and a pipe, who had a career in some qualitative discipline, a historian or an archaeologist.

Brad's an organic farmer, clean-shaven.

He and Jill visited for a couple days over the holidays. Our daughter drove in from school. Things were awkward, so small talk prevailed. Plus, there was wine.

I threw a party. They danced, Brad's chin on Jill's head. My friends peppered Brad with questions about his farming methods. Compost beats bone meal, he said.

The night before Brad and Jill were to return home, there were plans to meet for dinner with Brad's mother, who was flying in for the day. I made reservations at a Thai place.

That afternoon, I wandered through the house, looking for things of Jill's. When I drove into the city that night, I took what I had found: A copy of "The Female Eunuch" and an embroidered hankie I bought for her in Paris.

I stood in the lobby of the hotel when Jill came down to announce that Brad's mother had been taken ill, fluttery stomach, lightheadedness.

"She wants us to leave her and go to dinner. I don't think Brad wants to."

"His call," I said.

"Maybe next time," Jill said.

"I found these today," I said, holding out the book with the hankie on top. Jill grabbed the hankie and stuffed it into her blazer's breast pocket. "Keep the book," she said, "I never cared much for Greer ... sex-fixated, so boring."

With that, she spun around and headed to the elevators. I stood there for a bit, thought about getting some Duck Choo Chee on my own. I went home.

It's a small thing, I suppose, the hankie, but I couldn't shake the thought of it, an exquisite piece of our life together, dismissed with a swipe, a dry cleaner perhaps the next to touch it, admire it, to fold it and return it to the blazer's breast pocket, maybe with a corner peeking.

Camp David Mom
Debra Gipson

When my son told me he wanted to join the Army, I was taken aback. I had always assumed he would follow in his father's footsteps and become a serial entrepreneur. Instead, he chose a path that would leave footprints not by Allen Edmonds, but by combat boots. I urged him to become an officer, but he was determined to enlist. I will never forget the day he shared his decision with me. It was 2014, he enlisted the year I became a veteran.

I told him, "I am not like most mothers. Most would discourage you from joining or protest outside Camp David if something happened to you. But I am not like them. If you choose this path, you must understand that it comes with risks, and risks carry consequences, some intended, others not intended." He told me he understood.

We have always been simultaneously too much alike and too different. We spent much of the next few years or so not speaking. I do not remember what it was about. We go through this often enough for me to know that it will course correct when it is time. One thing I am certain of, is that it is never about a lack of love. We love each other fiercely. We argue that way, too. And then one day, the day will come when he needs his Mommy and I need my best friend.

It was during one of these valleys that I fell asleep on the living room floor with the television tuned to a news outlet—something I do not often do. I was jolted out of my sleep by an announcement about a bombing in Al Asad. It was not the bombing itself, but a deep, terrible feeling that something was wrong with the son I had not spoken to in years.

I called his father and asked when he last spoke with him. It was the middle of the night, and he called me "crazy." I told him if he did not find our son, I would define "crazy." Do not worry. It is not a threat. It is more of a war cry. Anyway, I hung up and made some calls. The people I called phoned other people, and eventually, I was told, "It isn't pretty, but he's alive." I did not care about "pretty;" I cared about alive.

I called his father back, "I found him. He is alive," I said.

"It was fifteen minutes ago," he announced.

"That is the power of crazy," I said, and hung up, wishing for a brief and fleeting moment that I had an old-school, wall-attached phone so I could experience the petty-joy-foolishness that can only come from slamming down the receiver.

A year went by, and still, I could not get that night off my mind. One day, I decided to Google my son's name. That night, the night I was jolted from my sleep and made those frantic calls, my son had just come off a 24-hour duty when he was tasked with helping evacuate the base's assets and personnel before the attack. It meant he would not receive the protections he provided others. He endured over 80 minutes of continuous bombing and eleven close-proximity missile impacts for which he was awarded the Purple Heart.

Months after this discovery, I was driving when my phone rang. I answered, and the voice on the other end was unmistakable. We talked, listened, reconciled, and before he hung up, he said, "I love you, Ma," and I told him I loved him, too.

We hung up, and I wept with joy before a realization set in. That speech I gave before my son enlisted was a ruse. I too, am a Camp David mom. I too, have been to combat. I am the most Camp David of all Camp David Moms.

Oh, and those valleys? We found a way to fill those in and smooth them out. We found a way to listen more and fight less. He is one of my favorite people and the person I admire most in the world.

Another Day in the Desert
Brent Herzberger

Eyes wedged shut from the exhausted sleep slowly peer open to another day.

The first sigh of breath stirs desert "moon dust" into a sparkling cloud hovering over my chest.

The fog of a dream still present yet fleets faintly from my mind, my eyes shut again.

Closed, my mind attempts to recapture those traces of the dreamt visages of home.

The sum of days has lost prominence, blurred together in the focus of many tasks amongst missions.

Occasionally, we recollect the count, elated so many have passed, yet saddened so many remain.

In automaton fashion we prepare for the day, the road, the mission, the patrol, another day's end.

Yet, the tedious repetitions never dull our vigilance, we are each other's only keepers here.

Vehicles engines rhythmically expel blue-gray enveloping clouds odorous of diesel.

Mental lists checked and verified are in a defined sequential order not to be found on paper.

Nods and gestures signal ready and simultaneously acknowledge our pledges of fealty to one another.

Bonds stronger than any word, pledge, or oath can suffer bind this small fraternity of warriors.

Traversing the black vein of asphalt road, eyes intently focus on every angle of the barren desert scape.

Entering remote towns makes our alertness palpable in comparison to the previous open spaces driven.

Robotic voices over the single radio speaker confirm and verify our passage through the village gauntlet.

Tension recedes slowly, not forgoing the remaining journey of the road still requiring our focused watch.

The blinding red-orange orb of the setting sun, flanked by slight purple hues, slips below the horizon as we enter our refuge watched by stern sentinels in towered sandbagged perches.

This guarded haven, a respite, allows our hyper-attentiveness to leave our bodies in a tidal rush, leaving a wake of exhaustion in body, in mind, and in soul.

Head counts showing no debit of starting numbers is a triumph over the wagers of fate and happenstance. Collectively, we commiserate at the end of another interminable day.

We each singly make our way to exhausted, dusty slumber and quietly hope for another sleep-envisioned visit of home to embrace dreamily before the reality of the next day awakens us again.

Chappie
Brent Herzberger

Up before the blistering sun breaks the horizon, our scheduled time of departure draws nears for us to traverse again the dangerous routes of the barren desert landscape. The Chaplain has arrived.

We gather at the lead vehicle, encircling our Chaplain as if to protect him in this hostile environment in which he walks unarmed but shielded with his faith, we are eager for his words of prayer upon us.

Always, he meets us prior to each departure, no matter the time, no matter the day, he asks a heavenly blessing be bestowed upon us to wear spiritual armor under our ceramic plates.

Quieted heads bow reverently, believers and those who may not, oblige this uniformed man of God as the faithful are nourished and the uncertain do not want any hazard left to chance.

He prays the same prayer each departure, memorized after countless combat patrols, all circled around him silently mouth each word in unison with him as he begs God's protection upon us in our tasks.

A choral "amen" signals time for departure has come, mounting vehicles and conducting last checks, each sounding off readiness and each waves to our chaplain as our convoy rolls out.

We drive out, again into the daily danger, emboldened and bound together by just that brief shared spiritual moment and confident we will see our next departure where our Chappie waits again.

Oh, So Long
Dennis J. Jackson

She waited for him at the airport
It had been just too long
Lonely days of silence
Thinking of their favorite song

He went to serve his country
The best that he could be
Standing up front for freedom
So that others could be free

High school sweethearts together
Married by a Justice of the Peace
He shipped out to basic training
Four years till his release

He then chose to go Airborne
Protectors of this great land
First to be activated
First to make a stand

With only a late-night phone call
Tears and I Love Yous
He promised to be careful
This is what he had to do

The letters were infrequent
She could only understand
Honoring his commitment
Serving in the burning sand
Then came the notification

The Chaplain at her door
Her loving soldier husband
Won't be coming home anymore

Today she's waiting at the airport
One red rose in hand
She's waited oh so long
To stand next to her man

The flag draped box rolls slowly
Funeral detail in dress greens
The tears flow so softly
It all seems just like a dream

She walks beside her husband
Now covered by the Flag
Life was once a fairytale
It's now become a drag

She holds onto his picture
So happy and full of life
So very proud to serve
So proud of his loving wife

Paid the ultimate price for freedom
For those he didn't even know
Such a loving, caring husband
That's why she loved him so

As Taps plays in the distance
And all the mourners cry
Some may soon forget
She will never really say goodbye

The Boys at the Wire
Douglas Kulow

There they stand waiting, staring up through the twists of barbed and razor wire. Saying nothing, they've learned that their suppliant faces work best with these Americans where they live. Unlike in the cities, where their cry "GI give me P?" might work within the hustle of commerce that includes anything and everything in both directions, Salem smokes and Tide for **cần sa** (a pack of joints) or something more, here the soldiers find their plaintive cries annoying. The boys know, if they bother one of *these* soldiers, he might throw something at them or spit at them or tell them to "f___ off." No, pleading words won't work with these GIs. So, watching and waiting and hoping, walking barefoot over the rocks and sand and street oil and ditch slime, they come in olive ball caps, and olive shirts and olive short pants. Where do they get those "OD Greens" in sizes this small? If they are old enough to even wear pants, that is.

As the story goes, there was a chaplain in I Corp that noticed how many children wore no pants at all. Recognizing that this nakedness was not at all in compliance with God's expressed intentions for mankind since the loss of Paradise, he wrote home to his congregation. That prompted them to take up a collection to buy several thousand pairs of children's shorts in red, white and blue. When they arrived in Vietnam, the chaplain ordered up a jeep and headed out to distribute this contribution of American modesty to the local villagers. He did not anticipate his inability to tell one small Vietnamese child from another as they accepted his gift, only to run behind the hooches, take off their shorts and run back for more. By the third village, his supply was exhausted and the next day there was a glut on the black market of red, white and blue children's short pants. Even when the fair market price for short pants bottomed out, sales were low. What he failed to consider is that with no bathrooms anywhere, there was a perfectly good reason the little toddlers are running around with no pants. Within a week, red, white and blue oil rags were showing up at Ernie Mayo's motor pool garage, just purchased on the black market in downtown Quy Nhon.

So, the boys at the wires stand and wait. So tiny, eyes so brown, so beautifully innocent. Plaintive and waiting for the gift of "P," piasters, the French name for Vietnamese money, or Dong, the Vietnamese name for the same Vietnamese money, or "MPC" Military Payroll Certificates, the money of the military in Vietnam. They don't care which. It all spends the same to *them* in an economy ravaged by this war and the money freely spent by *these* men that live on *this* side of this wire. The dollar for which a papa-san labors all day in the heat, filling *their* fortification sandbags, or for which a woman washes *their* clothes and cleans *their* barracks, that dollar is rendered valueless by the money these soldiers spend on one trip through the black market of Quy Nhon, in the shops and bars and brothels. Everything is bartered, everything is traded. The standard of exchange in the value of labor for currency is simple in 1967. It's a dollar a day in this combat zone. A dollar a day.

It's only been a couple years since Bob Dylan introduced us to the ironic concept of a dollar a day in lyrics from "Talkin' New York:"

Well, I got a harmonica job, begun to play.
Blowin' my lungs out for a dollar a day.
I blowed inside out and upside down,
The man there said he loved m' sound.
He was ravin' about how he loved m' sound…

A dollar a day's worth.

What is a day worth to these boys at the wire?
What great benefit has been brought by these soldiers to these little boys' lives?
Today, tomorrow, forever?
What promises do their lives hold now that we are here to fight for their future
freedoms?
Freedoms from what?
Freedom towards what? Education? Employment? Health?

And so today, these boys at the wire look up at us as we walk down the back stairs of our barracks. The bartering begins in pigeon English-Vienamese:

"**Bao nhiêu cho cần sa?**"　　　How much for some marijuana?
"**Năm đô la.**"　　　　　　　　Five dollars.
"No way. Too beaucoup. **Mot đô la.**"　One dollar.
"**Hai đô la.**"　　　　　　　　　The boy responds with two dollars.
"Okay. One for the **cần sa** and one for you."

You see it's a dollar a day's worth, here at the wire.

Entering this combat zone in 1967, our first learned words in Vietnamese included:

Dừng lại. Stop.
Nằm xuống. Lie down on the ground.
Cởi quần áo ra. Take off your clothes (to check for hidden weapons).
Xin lỗi. Sorry about that.
Chào cô gái. Hello (to a girl or woman).
Xin chào cậu bé. Hello (to a boy or man),
Anh yêu em nhiều lắm. I love you very much.

On the Other Hand
Don Lee

In August, Dad, Army/AF TSgt., kept the radar dome working during WWII on Attu Island, Alaska.

I was born, August 1944 in
Zanesville, Ohio.

I have uncles who served, the Army in WWII, around the Pacific Theater.

I tried staying in college for a draft deferment away from Viet Nam.

On summer break, I felt the draft and kept the grill hot and the dog sleds running in Fort Yukon, Alaska.

I'm proud of my son, the soldier, a career officer Marine.

Over There, Over There, I hear about and I'm wonderin' what tomorrow will bring.

Wherewithal, I Manage Again Gathering Inspiration, Not Expectations,

Wondering Where the Lions Are.

Movement to Contact: Day One.
Stephen May

The sun had just set, but the only way of knowing was that the grey sky was darker. We had been moving all day and we were tired. Earlier that day, all the non-combat vehicles had peeled off the seemingly endless sea of US military vehicles moving north, and now we were down to just armored fighting vehicles. The artillery tracks had stopped a short while ago and still our tanks and Bradleys moved forward. We had just moved through a barrage of incoming artillery fire that taught us all in short order why artillery is the "King of Battle." The rain of exploding steel claimed no victims but made us scramble to close our hatches. Each round that hit nearby rocked the 63-ton tank and made each of us wince with the blast. I had only a small periscope in the loader's hatch from which to view the devastation, but that tiny window was like a porthole into Hell. The sand, dust, rock, and smoke from each round blown into the air flew everywhere and great holes remained where they had hit. One of the platoon leaders protested over the radio about our slow pace and was quickly shut down by a curt response from the XO. We moved on.

We moved through the darkness in the eerie green glow of the IR viewer in the driver's hatch. The dome and instrument lights were dimmed in the turret, so as not to destroy our night vision and disrupt the IR goggles, with which I scanned from the open hatch. The landscape was barren and showed no sign of human life. As the tracks rumbled beneath us, the radio announced in our headphones that there were no friendly forces to our front. We were now weapons-free to engage any movement to our front. Shortly afterward, our gunner Sgt. Bosley alerted the crew to vehicles to our front. Lt. James peered into the TC's sight and confirmed what he had seen. Lt. James keyed the company net and reported what we had seen to the company commander and was given authorization to fire. I leaned over and switched the radio back to the platoon frequency and gave the LT the thumbs up. Lt. James announced to the platoon it was time to go to work and then turned his attention back to his crew. As Sgt. Bosley laid the reticle on the target, Lt. James issued the fire command: "GUNNER, SABOT, TRUCK!" The crew sprang into action as the mechanisms of destruction began to function.

Sgt. Bosley shouted "IDENTIFIED!" and switched the thermal sights to 10 power for fine adjustment to the gun lay. The whine of the turbine engines howled with the pitch of an enraged demon.

Simultaneously, I shouted "UP!" and moved the arming lever to the up position, thereby arming the main gun. The breach held a 120 mm Armor Piercing, Fin-Stabilized, Discarding Sabot round that had been riding patiently there for a week. We were told to battle carry Sabot before we even crossed into Iraq, as we needed to be prepared to kill tanks. It was overkill for the first target being a truck, but we would correct that with subsequent rounds.

I tensed up to reload and concentrated on the breach as Lt. James gave the command to fire. Sgt. Bosley hollered "ON THE WAY!" and pulled the trigger. The breach jumped and spat out the aft cap of the expended round. The clanging of the landing aft cap was still ringing in the turret as I kneed the switch that slid open the ammo compartment door to reveal the rows of dimly lit cannon rounds. The faint blue light from the dome lights helped me to identify the heavy black H I had written in marker to tell which rounds were High Explosive. I seized the lip of the round with one hand and tripped the extractor lever with the other. The long, black High Explosive Anti-Tank round slid smoothly from the storage honeycomb and into my arms. A quick flip end over end lined up the business end with the gaping hole in the breach and with one quick slam, it seated in place. The breech snapped shut and I screamed "HEAT, UP!" The whole process took about four seconds.

As I was going through the motions of reloading, the first round had found its target. The gunner and TC both let out a whoop as the splash of sparks downrange confirmed the kill. The scent of cordite wafted in the tendrils of smoke inside the dim turret. Sgt. Bosley leaned back into the gunner's sight to find another target. It wasn't long before he announced that he'd found one. The sequence repeated itself as Sgt. Bosley indexed HEAT so the ballistic computer knew the ammunition had changed. This time as the command to fire was given, I popped my head from the hatch to observe the impact. The cold night air was displaced as the heat and concussion burst from the end of the gun tube. The bright flash of flame took my night vision away and replaced it with the photographic image of the muzzle flash. Off in the distance, I saw another flash as the HEAT round detonated against the target. I popped back inside to feed the cannon and fumbled to find the ammo with my diminished night vision. Lt. James called for another HEAT round and as the ammo door slid open, I grabbed the first H I found. I stuffed the round into the breach, being careful not to bang the standoff spike against the inside of the turret. The Breach block hungrily gobbled up the round and I armed the cannon with a resounding "UP!"

We engaged three trucks and an armored personnel carrier that night. When it appeared that we had a lull in the fighting, Lt. James gave the order to shut down and take a break. Huffman, our driver, killed the engine, and the Banshee ceased its mechanical wail. With the engine silent, the ticking of the thermal sights and the high-pitched hum of the gyroscope in the turret seemed to get louder. The driver's hatch banged open, and Huff emerged to light up a cigarette. He hadn't left the driver's hatch in a day or two and sure looked like it. Sgt. Bosley climbed from the seat deep in the belly of the tank, passed me in the loader's hatch, and lit a smoke off the end of Huff's glowing cigarette. We gazed upon the sterile desert landscape that was lit up here and there by the fires of numerous burning Iraqi vehicles. Each dead vehicle lit up a broad circle around it like a funeral pyre. The faint breeze carried not only the sound of small arms ammo cooking off in the flaming remains of the wreck, but the smells of burning rubber, fuel, and flesh. The popping of the small arms ammo sounded like popcorn and added a grim festive sound to the atmosphere.

We glanced at each other's faces in the flickering fire light. We were all dirty. Our faces were lined with dust, covered in stubble, and what little hair we all had was matted with filth. The chemical protective suits we wore were stained with diesel fuel and grease, but the charcoal linings kept the smell of our unwashed bodies from our noses. We were a sight to behold as we silently watched the fires burn around us. Huff took a long drag on his cigarette and held it, then he flicked the glowing butt off into the void at the base of the tank. We drank warm, gritty water from our canteens as Sgt. Bosley relieved his bladder off the back deck into the sand. Lt. James's face was tired. The man had stayed awake for the past three days on adrenaline and force of will alone. But he finished his reports to higher then took a drink from his canteen. We had all aged years in the week since we crossed the border. We climbed back into our positions and the tank's engine slowly spun back to life. Huff dropped it into drive, and we crept forward into the darkness.

Vendetta
Raye Oheidhin

I am in a battle
A prolonged
Bitter quarrel
With PTSD
I fight
A brawl in my head
With the devil
I seek vengeance against
The illness that plagues me
I would welcome
A face-to-face meeting
Run-in
To face my worst fears
Rather than the
Hyper-vigilance
Vivid dreams
Flashbacks
That PTSD feeds me
I am a warrior
With a wish
To simply win the battle
The war in my head

What Color Is Your Love?
Kevin D. Smith

I got one question for you ...
What color is your love?
Is it red like the fire burning in a thousand suns? Hot to the touch but where the f_'s the fun?
What color is your love?
Is it cool, true and icy blue? A puzzle for someone who can and f_ing can't pick up your cues.
What color is your love?
Is it mean, lean and envious green? Eyeballin' everyone that makes the scene.
What color is your love?
Can it be that deep, dark midnight black? Swallows you whole and you never come back.
What color is your love?
That golden brown, that golden brown, that golden brown, sweet sticky molasses holding you down.
What color is your love?
Is it a yellow sunrise, tickling my face with light kisses greeting me with new chances?
Maybe orange like a summer setting sun? Memories of wine, laughter and slow dances?
What color is your love?
Maybe it's f_ing white? No delight. Blank and devoid.
No passion out of fashion you want to avoid.
So, tell me ...

What color is your love?

Invisible
Holly Spohn

The year prior to my deployment to Iraq, I had shot better than everybody on the .50 caliber machine gun, more endearingly referred to by most soldiers as "Ma Deuce." Maybe that's why they picked me when they needed somebody to man "The Pig," an M60, smaller caliber, machine gun. Prior to that moment, I had only ever seen "The Pig" in the old Rambo movies. You know, the scene in Rambo: First Blood Part II where he single-handedly shoots everything up with a fully automatic machine gun? That's the one. I'm pretty sure "The Pig" was already obsolete at the time my sergeant handed it to me. It was a Vietnam-era relic as far as I could tell. These weapons had been left behind by the prior unit and adopted by us when we took over.

As I took the weapon from my sergeant, I had only one question, "Open bolt or closed?" wondering what position I needed the bolt to be in for the weapon to be ready to fire. Once I got my answer, I positioned myself behind a concrete barrier looking west down a street that led to our position. I was pulling security for a civilian construction team just outside a gate. Hours went by before a taxicab came speeding around the corner in our direction. They seemed to be speeding up as they approached. I moved myself into a position to withstand the recoil I expected, suspecting the M60 should have probably been mounted, although we did not have a mount for it. I braced myself. I could hear the Albanian commando next to me yelling "No! No! No! No!" afraid that I was going to start shooting wildly without thinking.

Years later, I was at a bar and a couple of veterans were talking about their time in the military. I had seen them around before and even talked to them, but not about my military experiences. But on this day, one of them mentioned the M60 and I started to tell that story. They just talked right over me. I tried again during a pause in conversation, but they still would not let me join in. They didn't even look at me. It was as though I was invisible. To this day, I cannot really understand why. I suppose I never will. I can only piece my life experiences and the experiences of others together to try to understand. What those experiences have shown me is that it is because of what I am. Because I am a woman, my story takes something away from theirs. A woman talking about manning an M60 in Iraq made their stories somehow seem less valuable.

Even Uncle Sam treated me differently because of my sex. Before I even signed up, he told me I was different. I was on a school field trip at the Ravenna Arsenal when I decided I wanted to join the Army. Soldiers there took us out on the various tracked vehicles. We rode an M113 Armored Personnel Carrier and the M1 Abrams Tank. It was after that ride on the tank that I decided I wanted to join the Army. I was going to be a tanker. When I said it out loud, the staff sergeant driving the tank said to me "You can't. You're a girl." I didn't understand. "What's so hard about driving a tank?" I shot back. He told me females are not allowed to serve in combat arms specialties and a tanker was just that, combat arms. At 18 years old, as a girl who had been raised to believe she could do anything she wanted, that was the dumbest thing I had ever heard. But I still joined. And I picked the next best job specialty. It was a specialty even a girl could do.

I became a tracked vehicle repairer, or as I tell my civilian friends, I was a tank mechanic. Why did I choose that job? Because, as a tracked vehicle repairer, I could drive tanks. That's really what I wanted to do. When I got to my unit, after my initial entry training, my team never treated me differently because of my sex. I was one of the guys. I did the same things they did. I crawled under vehicles, I pulled power packs, and, of course, I drove the tanks. I had all but forgotten that I was limited by my condition of being female. I was fortunate that I rarely felt different when I was with my team. Sure, there were times some guys treated me differently, but not my guys. Among the guys in my section, I was one of them. It was not until after my service ended that I really started to feel different.

Now, I am continually reminded that I am not like my brothers. When I get stopped by the police and the officer sees my veteran plates, he asks me if my husband served. I'm not married. And my cars are registered in my name. I always smile and politely say "No sir, it was me." He thanks me for my service and lets me go on my way.

Like many of my military sisters, through these experiences, I have learned to quietly fade into the background and keep my service to myself. I don't go to veteran events and talk with the guys about my service. Whether it is true or not, I always know that my service is less than theirs. If I see females at a veteran event, I will likely try to connect with them, because I know I am like them. Unfortunately, women frequently don't attend veterans' events in the mass numbers that men do. Many of us do not attend because we have found we do not belong.

I sign up for veteran events, seeking the camaraderie I once knew when I was in the military. When I get there, I am reminded I am not one of them. Just recently, I was participating in a veteran hike. I volunteered to help with the event. But, once again, I seemed to be invisible among my many brothers and the few sisters who came out. During the event, we stopped at the veteran's memorial where a man approached the group. He walked through the crowd of veterans, fist-bumping each one along the way and thanking them for their service. He walked right up to my buddy who was standing next to me, gave him a fist-bump, thanked him, and walked on past me to the next male veteran in the group. Over time, experiences like this have changed the way I view my own military service.

One year, my friend's son invited me to a Veterans' Day event at his elementary school. When they asked all the veterans to stand, I remained seated. My friend later told me her son asked why I didn't stand up. I didn't have an answer. Back then, I didn't really know why. Looking back with a clearer perspective of my experiences and having found a female veteran group where I feel like I belong, I now know that was another instance of me just fading into the background. I had internalized what my experiences had been telling me—my service does not count. I am not worthy to stand among these men who served their country. I am not a real veteran.

Because I am a woman, I do not feel that I can be proud of my service, or if I am, I should keep it myself. We do not all feel this way, but I know many of us do. You won't often see my sisters or me walking around the grocery store in our combat veteran hats. You might see me in an Army sweatshirt on occasions, but I expect you will assume it is not mine. If you ever see us among a group of veterans, you might not even notice us. We don't stand out. We just fade into the background. We often won't approach our male brethren, because we have learned through our experiences that we may not be accepted.

The funny part of it is, I never really knew what was happening. I never knew I was fading into the background. I did not know why I didn't stand at that Veteran's Day event. I only knew it didn't feel right for me to do so. It was not until I found a group of female veterans just like me, that I started to feel free. My female veterans' group has shown me that I should be proud of my service because I see my veteran sisters and see that they should be proud of theirs. It wasn't until I found my place that I felt comfortable standing among my brothers. It was not until I felt like I belonged that I began to understand why I always felt like I didn't.

Gratitude is not what I seek. What I want is to connect on a level that soldiers connect on. What I want is to feel that bond I once had with my brothers and sisters in the military—an inherent understanding that I could count on the soldier to my right or my left no matter what. It is a bond that is like no other, an organic and wholesome relationship that needs no nourishment to survive. Veterans can meet for the very first time and instantly connect simply because they are veterans. But all too often, female veterans don't get these opportunities, simply because they are female. We struggle to even connect with other female veterans because so many of us are fading into the background, taking the place we believe to be ours. We are not only invisible to you, but we are also invisible to one another. No, gratitude is not what I seek. What I seek is to be seen. What I seek is to be accepted. What I need is to know that I am not alone.

So, what I ask of you, if you are a veteran and you see me among your group, is to say "Hello" and reminisce with me about our shared service. If you are a female veteran, fading into the background and feeling like you do not belong, come join me and make your presence known. And if you are an organization that serves veterans, think about me, and know that I may not be at your event because I am invisible.

We'll Always Have Paris
Jaelle Terrell

Victoria stretched out, groping for the phone on her nightstand. "What time is it?" The pitch-black room was perfectly still. She hadn't used a clock in her bedroom since her last heart attack. Victoria hadn't woken to an alarm clock since her retirement. There was no reason. "4:30. No time for a civilized girl to get up." Victoria let her mind roam through the seventy-five years of her life. She enjoyed the memories. Her first kiss, for example, was a particularly fond one.

She was fourteen or fifteen, and quite tall. It was a winter evening, and it was snowing. A light snow, with those giant snowflakes that seem to float down from the sky like feathers. She can still remember those big snowflakes landing on her face and the face of Karen. Yes, Karen. Victoria wasn't always Victoria.

She was born James. Jim was a skinny boy, tall and gangly, and pretty severe allergies had left him slight and not very athletic. He was an avid reader who loved to watch old movies with his mother. This exact memory is that Jim and Karen laughed as they negotiated the difference in their heights, and Karen stepped up onto a porch step and turned her face up to his. Jim thrilled as his lips met those of another person, with those big, beautiful snowflakes falling on both of them. "What a wonderful memory," Victoria thought.

'I wonder if Karen is out there. If she remembers it."

But that was not this morning's reverie. This morning she laid in bed recalling when she was an eighteen-year-old boy newly enlisted in the U. S. Navy. She was stationed in San Diego in Amphibious Construction Battalion One. The Seabees.

Jim was eighteen in 1981. There was no such thing as transgender, in his world anyway. Jim loved girls, and the Beatles, and pot. He had gravitated to a group of similar boys, and they would buy their weed for $15 an ounce. They lived in modern barracks, four to a room. Someone had found a rental house in a neighboring community, and they were all going to go in together to rent it and live off-base. It was all set and happening soon.

Jim was given a temporary assignment when he first reported to ACB-1 before his permanent position became available. It consisted of running a floor buffer through an empty barracks about once a week and getting high and playing cards the rest of the time.

One day, Jim was pushing the buffer through the barracks when he realized someone was standing in front of him. He shut off his machine and looked up to see another boy dressed in the same uniform. "My name's Chris," he said with a smile. "I got a room here and one of my roommates is getting out and going home." Jim looked Chris up and down. He was about 5'10", with blonde hair and big perky brown eyes, and quite good looking. "I don't know, I wonder if you want to move in?" Jim thought about his plans with his friends and the house they were poised to move into. Of course, he said yes.

Victoria chuckled to herself. Jim never understood why, in that moment, his life took such an immediate and dramatic turn. But Victoria did. Chris was handsome. Beautiful. And, Victoria supposed, it was love at first sight.

Jim broke the news to his friends. They responded with, "Huh?" "We agreed!" "Why?" and every other logical question one could imagine. But Jim had no logical answer or rational response. He just stuttered and stammered and hemmed and hawed. And he moved in with Chris.

This began an absolute highlight of Jim's life. Yes, of Victoria's life. For one thing, Jim was born in Cleveland, Ohio. Cleveland was dirty, the people were hard, the river had caught on fire. Cleveland was like Chicago's cousin. If there was a rivalry, a connection between them, Cleveland was the cousin you only see at holidays and funerals. Chicago had a class, a culture and sophistication that Cleveland didn't seem to possess. It could be worse. Cleveland could be Youngstown or Scranton. Chris was from Portland, Oregon, the polar opposite of Cleveland. Portland was polite, friendly and clean.

They were together for about two years, strictly in the platonic sense. They were joined at the hip. They were two peas in a pod. But neither of them ever pursued, or even talked about their closeness. They never shared a word, a touch, or a caress. But there was a bond, a bromance, yes, a love between them.

Victoria lay there, reflecting on Chris. "He gave me so much," she thought. Chris gave Jim music. For the first time Jim heard Grateful Dead, Poco, country rock and surf music. And much more. He gave Jim books. For the first time, Jim read Tolkien, Hesse, and Huxley. And much more.

They put down an area rug in their barracks room, added a sofa and a hanging lamp and made it a comfortable living space. Everyone who visited them said it made them feel like home. How many evenings they sat on that couch, smoking dope, and sharing thoughts and dreams! And Chris would draw. He could draw beautifully. He would doodle little things or draw big, complex things. Then he would wad them up and discard them. Jim loved everything Chris drew. In fact, Victoria still had a piece of Chris' art somewhere packed away.

They would go out and explore San Diego. They went to the beaches and found Black's Beach. It's a nude beach, and Jim just walked out and stripped. Chris went behind a bush to disrobe, and then emerged nude. Jim didn't understand that at all. Today, Victoria finds that modesty charming. They went to concerts, usually at the San Diego Sports Arena. They often went to Balboa Park. It's a lovely place filled with music and art, friendly people, and a museum or two. And they went to the zoo. The San Diego Zoo was the most beautiful place Jim had ever visited, and they spent many days there, visiting the animals or just sitting in one of the three giant aviaries.

There were times, perhaps many times that they argued or fought. This morning Victoria was sure the fault was usually Jim's. Jim could be petulant, a brat, and Chris always called him on it.

Once, Chris was named Seabee of the Month. He was neat and clean. His uniform was always sharp and crisp. He was polite and did his work promptly and thoroughly. Of course, he was named Seabee of the Month. This got your name in lights. You had a special parking space, a plaque, and a couple more little perks. Yes, Jim was a little envious. It was a nice honor. Chris stood up, looked the captain in the eye, and said, "No thank you. I am a conscientious objector. I don't believe in the Navy and I don't believe in the Vietnam War. So, it would be hypocritical, it would be wrong to accept this award." Victoria was sure no one had ever done such a thing before, and perhaps since.

One day, Chris' enlistment was up, and it was time to go home to Portland. Their time together was over. Victoria remembered a moment together, a goodbye with mild assertions of affection, and then he was gone.

But the changes in Jim, and Victoria, are measurable. He was nicer, cleaner, and smarter. He was a better person. Jim carried such fond, wonderful memories of his time with Chris. Victoria is not good with computers or the interweb, and she has searched for Chris to no avail. She has wondered if Chris is still alive. "Did he come out as a gay man? Or did he marry and have 2.5 children?" She has no regrets about not pursuing a more intimate relationship, because those two innocent boys have provided a multitude of warm, happy memories. Besides, an unwanted advance might have resulted in a rift or an end to the relationship they did have. Besides, they were both so naïve.

It wasn't until years later that Victoria came to understand and embrace her own gender identity. "I wonder," she thought, "Would Chris recognize me today? Would he accept me? Would he run away in terror? Ahh," she sighed. "I guess I'll never know. But like Bogey said in Casablanca, we'll always have Paris." She threw off the blanket and headed to the coffee pot.

The Courage of Spirit
Callie Travers

The story of the female soldier is seldom told.

From ancient times, the very first war, we have been there. Whether through mass
misinformation, misrepresentation, or plain evil, hidden under the guise of order, our
stories were lost. It is to our shame these stories have not been recovered.

The stories that do remain, rather slim, are at the risk of being tainted by their narration, their intention to form ideals. It is to our societal development that we uncover these hidden stories, as proof of our existence, long before the records show, long before we were included into their "Boys Club."

Women in the military, a hot topic even today, should be at best warm to cooling? Misogyny and fear persist, and our truth eludes us.

Through my experience as a female in the United States Armed Forces, I assure you my kind is but tolerated. Treated from the start as a problem in need of solving, or a secret in desperate need of a hiding place; I learned rather quick.

Learned to fall in, and quicker yet, I understood I would not be amongst the favorites, nor did I have any interest in earning that right.

The sentence that pervades my memories of my service time, like a

bell that refuses to quit
ringing, is the sentence that drives my hand.

"Don't be a Jessica Lynch!" had assaulted my ears from basic training in Fort Jackson, South Carolina, to the box training in Fort Polk, Louisiana, before deployment. An annoyance at first, feeling wrong for these sergeants (always men) to abuse this female and her name in jest, it turned into seething anger as the real truth came to light. I felt a set-up from the get-go.

Specialist Lynch survived an ambush in the Battle of Nasiriyah, in Iraq, on March 23, 2003.
She was attacked, kidnapped, and rescued in a true tale of today's war. I think of the countless times this exact scenario has played out in America's invasion of the Middle East. And of Lynch's many male counterparts who experienced the same crucial, chaotic, moments, the number of females for that matter, females who looked nothing like Specialist Lynch. There would be no cover of the *Army Times* for those soldiers, no media coverage at all.

It is my understanding that Lynch became stuck in her vehicle that fateful day in Iraq. From that
instant, her shock, her frantic, terrified moments, a new movement would form.

Instead of honoring Specialist Lynch for her service, her story became twisted into a weapon, fully loaded.

Judgement reigned supreme, and the message that at first felt so inspired, turned into one of "who not to be." We are so much more than a pretty face.

Specialist Lynch's actions, and reactions, the day her convoy was attacked may not have been on the side of heroic; it seems they were merely human.

Do we need to be heroes in order to be accepted?

Or, when we fit the mold they choose to create, are we set up like some sacrificial lamb in need of an altar? It's an atrocity that drill sergeants, any sergeant, could turn a soldier's worst nightmare into fodder to use on junior personnel. Lynch herself never portrayed herself as a hero, and on several occasions tried to set the record straight.

Yet, for this veteran, her name lists on repeat in my head. I'm proud to be like Specialist Lynch. Not only are we both light skinned, light eye/hair, supply specialists that served their country in time of war, but because we're both fighters, and most importantly, survivors.

Through stronger awareness and through our truth, women in the military may be seen for what strength they bring, rather than take away, of what we can uncover, not hide away.

I enlisted in the United States Army to protect my country, my own, no more, no less, than the man that stood next to me. In the face of insurmountable misrepresentation, I stand in the warmth and long for the cooling.

I wonder what name we could give to our female heroes? And, sorry, "heroine," although an
answer, is not in the running. That word is a whole other Pandora's box in need of unlocking. I think it's safe to agree, it is not a word one relishes attachment to.

Maybe it's a fact of manifestation, that once we envision a world where this word exists, then we can finally be known for who we are, rather than what you see.

Frosted Glass
Vickie Williams

It was a cold December night at Ft. Campbell, Kentucky, in 1986. No, actually it was cold as hell. So cold, car windshields froze in a matter of seconds. We – Sheila, Paulette, and myself – really didn't care how cold it was. We were soldiers and were tough enough to brave it. We got up at four-thirty that morning for PT, then worked all day. Now, it was time to hit the clubs in Clarksville, Nashville, or wherever. We traded our camouflage and combat boots for mini-skirts and high heels.

Jackson, a specialist from another unit agreed to be our designated driver. He was a quiet young soldier who played by the book.

Jackson drove us to a club outside of Nashville. Sheila, Paulette and I shooed him from our table because we wanted to meet other men and didn't want him to be a deterrent. He spent his time at the club socializing with whoever he found interest in.

Each of us met someone new that night. I had several dances with a guy named William Davis. In between dancing, we talked about ourselves, where we had basic, AIT, and where we grew up. I don't remember much about him except his sweater. It was an Argyle sweater. The same type of sweaters Bill Cosby used to wear on his television show, "The Cosby Show." It really stood out on him, the pattern and the colors: blue, brown, yellow and red all interwoven into each other.

After the club closed, around two-thirty, we all talked for a while in the icy parking lot while our vehicles warmed up. There were eight of us, four in the vehicle I was in and four in the vehicle

William rode in.

William and his three friends got into their car. Our windows were still fogged, so I could not see who got in the driver's seat. Our driver, Jackson, did not pull off right away. He waited until the vehicle defrosted and warmed up. Once we were able to pull off, we headed toward Ft. Campbell, KY, via Old Clarksville Road.

Many miles later, Paulette spotted a man staggering down the dimly lit road then shouted, "Vickie is that the guy you were talking to at the club?"

I turned my head to look out the window, "Yes, that's him. I can tell by his sweater!"

We pulled over and offered him assistance. He was disoriented and caked in blood with fragments of broken glass in his face, neck, and chest. He told us they had a car accident, and his three friends were injured, and he was walking to find help. We crammed him into the back seat of our vehicle. Jackson turned around and headed in the opposite direction Davis pointed.

Jackson drove slowly until he spotted a badly damaged vehicle in an open field. From a distance, I could see it was banged up beyond repair. Davis confirmed my observation just by saying the car skid on a patch of ice then rolled over a couple of times.

Jackson stepped out of the car and told all of us ladies not to get out because it might be too much for a lady to see. We shouted, "We're soldiers!" and got out anyway.

The accident scene was eerie: dark, smoky, and bloody. Davis's friends were in really bad shape. One guy was walking aimlessly close to the vehicle, discombobulated and screaming out for mercy. The second guys' legs were pinned under the steering wheel, and he was pleading to God to let him live because he had a kid to take care of. His body was twisted, and he faced the third unfortunate victim, the one who shocked me the most. His trembling and headless body was angled in the back seat. His head lifelessly lay in the open crushed trunk where it thrust through the back window. It was in spilled powder detergent and other whatnots. His eyes were bulged and appeared to be looking right at me.

Jackson was right. I may have been a soldier, tough and trained to kill, but I was a lady first, sensitive, caring, and the vessel for life. It deeply affected me. I had nightmares and flashbacks about his head. The thought of Davis's bloody glass-ridden face runs through my mind when I'm on dimly lit country-like roads or when I see Argyle sweaters. Memories of the accident victims trigger at night and also when I see powder detergent and messy trunks.

The windows may have all been frosted that cold December night, but the loss of life was very clear.

Thank You for Your Service, Fool!
Ken Woolfolk

After not sleeping and being up all night, I am fighting to stay awake by drinking chuckwagon, coffee, and watching reruns of Doc Martin on BBC TV. Were you ever thoughtful? Did you ever ponder?

Just maybe, you should have been more decisive, and less deceptive. Keeping it real with oneself, and not come across as a flake. You have the audacity to place a dime-size chip on your shoulder! That is not worth two cents! You spent too much time living in a fake-ass lavish life. The Fresh Prince of Saint Clair.

In Glenville Heights.

When you are expected to plan your life, all you have said is: My wife! My wife! My wife! They chopped one foot off and now they are eyeballing the other. I do not know why you are still mad at mommy and daddy; they have been gone for so long. I do not know why you still bother. Your stubbornness had you winning first place, the best dressed in a clown's attire. In all my years, thirty-five to be exact. In dealing with the Veteran's Administration. During which I have given them my blood, my sweat, my tears. I even trusted them to cut on me a few times prior. What gets me is that you never used your GI Bill, after 50 years, let alone using the services of the VA hospital. You are a deer in headlights, Bro. The way you sat there looking like an alive cadaver pepperoni pizza to them, two young newcomer residents who performed your amputation. They only stopped by to say, "Hi there! See you next week! We are taking the leg up to the knee!"

Thank you for your service, Fool!

The Butcher Brigade! You have been in the VA system six months or so, and you already have lost the limb. The family has been waiting for a lifetime, for you to stand up, to step up and step out on faith. When all you have done is sit back and play bridge. Thought you had everything under control, while the enemy was right under your nose. They are too close for comfort, and you do not even sneeze! That person with your medical guardianship, made damn sure they followed your death request to the tee. Making sure they do not catch a felony from nursing an incompetent so-and-so. Named Fool! It is a miracle that you did not expire from those low doses of arsenic that they kept reissuing your tastebuds. Now you are coughing because you are feeling a cold breeze. Giving the adversary goosebumps, butterflies, hot flashes, excitement, and cheap thrills. They place bets on when you will have your last meal! There is nothing I can do! This is a predicament that I am staying far, far away from. These are the beats from your drum, and it is defective. Wave the white flag, blow the foghorn! It is the only chance that you can get some peace.

Thank you for your service, Fool!

Contributors

Richard Asbury is a playwright, author, and retired educator (MS Ed. and MA Ed.). He was a commissioned US Naval officer for 21 years, specializing in Surface Warfare.

Kevin Chapman is a US Navy Persian Gulf combat vet, Acoustic Neuroma survivor, and writer. He has run the Cleveland Eastside Writer's Group for over fifteen years. He lives in Mentor, Ohio with his wife of thirty-two years, Teresa. Rounding that out are his two sons, Jacob and Nicholas, who are sick of his Dad jokes.

Michael T. Conway, Esq. is a former Marine Corps Infantry Captain who served with the 9th and 25th Marine regiments. He practices trial law in Ohio and Texas and is a member of the Bar of the US Supreme Court. He is married, has a daughter in college, and lives near Cleveland, Ohio.

Cheryl Darby is a veteran of the Unites States Navy.

Paul Facinelli is a retired journalist/teacher living in Avon with his cat, Ophelia. His work has been published here and abroad. He is at work on a children's book. Facinelli served in the Navy from 1964-1968, with assignments at Naval Air Station, Whidbey Island, Wash., and Kadena Air Base, Okinawa, where he was a copywriter for Armed Forces Radio & Television.

Debra Gipson, an Army veteran, is a screenwriter, podcaster, and filmmaker known for her engaging war stories that blend humor with compelling characters. A friend of Literary Cleveland, she is currently interning with the Greater Cleveland Film Commission while working on her fifth script titled "Swing," which focuses on a Negro League baseball team during World War II.

Brent Herzberger served for over 33 years in the active Army and Ohio Army National Guard. He retired in 2023 and now enjoys spending time with his family and his two grandchildren. He is an active member/officer of Hubbard VFW Post 3767. He enjoys being able to share his military experiences through writing.

Dennis J. Jackson is a veteran of the US Army 82nd Airborne Division.

Douglas Kulow served in the US Army from 1966-69.

Don Lee is a veteran of the United States Air Force.

Stephen May joined the Army in 1987. He served four years in active duty, including during Desert Storm with B Company, 4/64 Armor. He also served nine years in the Ohio Army National Guard. He currently lives in Columbiana, Ohio with his wife Caroline and their two dachshunds.

Raye Oheidhin is a former U.S. Army Sergeant First Class. She is a disabled veteran with PTSD who resides in Malvern, Ohio. Raye is the mother of two grown children and teaches kindergarten through fifth graders that struggle learning basic literacy skills. She's an avid reader, writer, and painter.

Kevin D. Smith is a native Clevelander who served in the U.S. Navy from 1981 thru 1985. He is the pesky big brother of two sisters, father of one daughter, grandfather of four, and great-grandfather of one. Kevin is also the band leader and guitarist of Xcetera, a local cover band that performs around the city. He is proud to now add "published poet" to his biography.

Holly Spohn served as a Tracked Vehicle Repairer in the Ohio Army National Guard from 2001 to 2013. She deployed to Mosul, Iraq in 2006 as a Sergeant of the Guard, performing base security operations. Holly now serves her community at home as an Assistant Prosecuting Attorney. She is a proud member of VFW Post 1055 in Ravenna, Ohio and Clearview HOPE Female Veterans Golf Group in East Canton, Ohio.

Jaelle Terrell is a veteran of the United States Navy.

Callie Travers lives in Cleveland, Ohio. She is the mother to three beautiful children and an Army veteran, using her writing to explore her past military experience with an emphasis on the experience of women who have served their country as soldiers and the trauma that remains.

Vickie L. Williams is a playwright from northeast Ohio. Her plays were staged for readings, workshops, and productions at Cleveland local theatres, in New York, and in Cambridge, Great Britain. She is a Puffin Foundation and an Ohio Arts Council Individual Excellence Award recipient. Vickie has earned a B.A. in communications and has contributed works to several anthologies.

Ken Woolfolk served in both the United States Army and United States Navy.

Acknowledgements

Thank you to poet and veteran Mansa L. Bey who designed and led the Veterans' Voices project. Thanks to Christopher Johnston who led workshops for veterans who do not wish to write about their military service. Thanks to Program Director Michelle Smith and Executive Director Matt Weinkam at Literary Cleveland for coordinating the program, and to Development Director Ryan Kearns for writing the grants that funded the project. And deepest gratitude to Rosalie Diaz, Deborah Derflinger, Jeremy Streem and everyone at the VA Northeast Ohio Healthcare System Whole Health group for tireless coordination, collaboration, and encouragement. This project would not be possible without their support. And our thanks and gratitude to the Char and Chuck Fowler Family Foundation and the National Endowment for the Arts Creative Forces grant through the Mid-America Arts Alliance for funding this project.

About Literary Cleveland

Literary Cleveland is a nonprofit organization and creative writing center that empowers people to explore other voices and discover their own. Through an expanding roster of multi-level classes, workshops and events, Literary Cleveland assists writers and readers at all stages of development, promotes new and existing literature of the highest quality, and advances Northeast Ohio as a vital center of diverse voices and visions. For more information, visit www.litcleveland.org.

The Ghostwriter's Wife

A Novel

Douglas Debelak

The Ghostwriter Series Book Two